# COMPETITIVE PHYSICS

## INTRODUCTION

This objective physics series provides a basic and challenging problem of physics from particular topics. It can be used to brush up ones basics and checking up the preparation level of particular topics. It is equally helpful to the traditional classes as well as competitions. It can be also taken as a revision material for any competition which includes the test of basic physics. If you want to grasp the subject before practicing these multiple choice questions, you can go through the website http://www.ncert.nic.in/ncerts/textbook/textbook.htm and down load the free copy of science books and after having command on the topic practice it. For revision purpose, important points are given at the starting of each topic.

# CONTENTS

# 1. MOTION

## SOME IMPORTANT POINTS

➢ When an object changes it position with time that would be motion.

➢ The origin is the reference point which use take to describe position of an object.

➢ The total path covered by an object is its direction.

➢ Magnitude is the numerical value of a physical quantity.

➢ The shortest distance covered from initial to the final position an object is called its displacement.

➢ The magnitude of displacement may be zero some time.

➢ When an object covers equal distance in equal intervals of time, it is said to be in uniform motion.

➢ The speed of an object is the distance covered by the object per unit time.

➢ The speed of an object with direction is called velocity.

➢ The SI unit of speed or velocity is m/s.

➢ The rate of change of velocity of an object per unit time is called acceleration.

➢ The Sin unit of acceleration is $m/s^{-2}$.

➢ The circular motion is an example of accelerated motion.

➢ The SI unit of distance and displacement is metres.

➢ When an object moving with uniform speed in distance – time graph, a straight line inclined at $45^0$ with time.

➢ When a line parallel to time axis in a velocity time graph it is said to be in uniform motion.

➢ The equation of motion in which v = final velocity u initial velocity, a = acceleration, s = displacement.

➢ $v = u + at$, $s = ut + 1/2at^2$, $2as = v^2 - u^2$.

# 1. MOTION

1. An object is said to be in motion when:

   a. it changes its size

   b. it changes its state

   c. it changes its position with time

   d. it changes its direction

2. To describe the position of an object we specify:

   a. a reference point

   b. direction

   c. time

   d. none of these

3. The numerical value of a physical quantity is its:

   a. size

   b. state

   c. magnitude

   d. none of these

4. The SI unit of distance is?

   a. meter

   b. meter/second

   c. kg/second

   d. km/second

5. An object is said to be in uniform motion when it covers:

   a. equal distance in unequal interval of time.

   b. equal distance in equal interval of time.

   c. unequal distance in equal interval of time.

   d. none of these.

6. Meter per second is the SI unit of:

   a. distance

   b. displacement

   c. speed and velocity

   d. none of these

7. To specify the speed of an object we only require its:

a. magnitude          b. SI unit

c. direction          d. none of these

8. The speed of an object is a:

   a. scalar          b. vector

   c. none of these   d. both a and b

9. The average speed of an object is obtained by:

   a. total distance/total time taken

   b. total displacement/ total time taken

   c. total speed/ total time taken

   d. none of these

10. An object covers 20 km in 2000 seconds and the other 30 km in 2500 seconds. What is the average speed of the object?

    a. 5m/s          b. 8m/s

    c.11.1m/s        d. none of these

11. When an object has speed with direction it shows:

    a. velocity      b. acceleration

    c. displacement  d. none of these

12. The distance covered by any object can be calculated by a device called:

    a. speedometer        b. odometer

    c. thermometer        d. none of these

13. The velocity of an object can be changed by changing:

    a. the object's magnitude       b. direction of motion

c. both a and b            d. none of these

14. The velocity of an object remains constant with time in:

     a. uniform motion            b. non uniform motion

     c. both a and b            d. none of these

15. When an object is moving with changing velocity, the object is said to be in:

     a. uniform motion            b. non uniform motion

     c. both a and b            d. none of these

16. The SI unit of acceleration is:

     a. m/s            b. $m/s^2$

     c. $kg/s^2$            d. m

17. When acceleration of an object is opposite to the direction of its velocity, the acceleration is taken as:

     a. positive            b. negative

     c. neutral            d. none of these

18. When a body falling freely from a height it is said to be in:

     a. uniformly accelerated motion     b. non uniformly accelerated

     c. none of these            d. both a and b

19. A car decreases its velocity from 90m/s to 30m/s in 10s. Find the acceleration of the car?

     a. 6m/s            b.-6m/s

     c. 8m/s            d. 3m/s

20. The acceleration of an object is a :

a. scalar quantity        b. vector quantity

c. independent        d. None of these

21. Which independent quantity is taken as a reference in any graph?

a. speed        b. Time

c. distance        d. none

22. In distance time graph which term is taken along the y-axis?

a. time        b. distance

c. velocity        d. none of these

23. When the distance time graph is straight line inclined at 45 degree with time axis it shows:

a. uniform motion        b. non uniform motion

c. object at rest        d. none of these

24. When the distance time graph is straight line parallel to time axis, it shows object is in:

a. non uniform motion        b. at rest

c. uniform motion        d. none of these

25. When the speed time graph is a straight line parallel to time axis, it shows:

a. none uniform motion        b. uniform

c. rest        d. none of these

26. When a car moving with uniform acceleration shows, velocity- time graph is:

a. a straight line inclined at 45 degree with time axis

b. a straight line parallel to time axis

c. a straight line parallel to velocity axis

d. none of these

27. When a body runs in a rectangular track once, how many times it changes its direction:

a. 1                         b. 4

c. 3                         d. 5

28. When an object moves with a uniform speed in a circular path it shows:

a. uniform circular velocity          b. non uniform motion

c. objects at rest                    d. none

29. The SI unit of deceleration is:

a. m                         b. m/s

c. $m/s^2$                   d. $-m/s^2$

30. The rate of change of velocity is:

a. acceleration              b. speed

c. distance                  d. force

31. the initial velocity of a body is u with an uniform acceleration a, its distance s at any time t is given by:

a. s=v                       b. $s=ut+.5at^2$

c. $s=ut+at^2$               d. none

32. a car with speed 850m/s in 1800 second, travels a distance of

a.1530000m                   b.114500m

c. 14000m                    d.15000m

33. A bus increases its speed from 10km/h to 20km/h in 5 second. What is its acceleration?

   a. 0.55m/s²           b. 5m/s²

   c. 4m/s²              d. 3m/s²

34. A car starts from rest increases its velocity by 20km/h, its initial velocity is :

   a. u=20              b. u=0

   c. v=0               d. none

35. When a body falls freely, what changes rapidly :

   a. distance          b. time

   c. velocity          d. force

36. Motion of moon around earth is an example of:

   a. uniform motion        b. accelerated motion

   c. non uniform motion    d. none

37. Which statement is incorrect for distance?

   a. it can be zero         b. it is a vector quantity

   c. it can be negative     d. all of the above

38. When a body shows uniform motion its acceleration is:

   a. zero              b. more than zero

   c. negative          d. none

39. Which formula shows relation between distance s, speed v, and time t?

   a. f=ma              b. a=(v-u)/t

   c. v=s/t             d. none

40. When a body is moving with a variable acceleration, the velocity time graph is:

    a. curved                b. straight

    c. none                  d. both a and b

41. The area enclosed by velocity time graph and the time axis is the:

    a. distance travelled by object     b. speed covered by object

    c. none                              d. both a and b

42. When final position of object coincides with initial position, the displacement will be:

    a. zero              b. increases

    c. at rest           d. none of these

43. Distance of an object can be:

    a. less than displacement        b. not less than displacement

    c. none of these                 d. both a and b

44. The acceleration of object is in the direction of velocity when:

    a. u is greater than v        b. V is greater than u

    c. a = positive               d. both b and c

45. When odometer of an automobile shows increment in its reading it shows:

    a. distance increased         b. speed increased

    c. time increased             d. none

46. A graph is always plotted between:

    a. Two variable quantities        b. two same quantities

c. Three variable quantities          d. none

47.    A car starts from a point A and reach to the point B. after some time car reach back to the point A. the displacement of the car is:

a. constant                b. zero

c. One                     d. none

48.    The formula of acceleration of an object is:

a. a = (v-u)/t             b. a = ut+.5at^2

c. a = v                   d. a = v+t^2

49.    Which is not an equation of motion?

a. v = u+at                b. f= mg

c. s = ut+.5at^2           d. all of these

50.    An object starts from rest increases its velocity by 20km/h in 40 second. Its acceleration will be:

a. 0.14m/s^2               b. 2m/s^2

c. 5 m/s^2                 d.8m/s^2

**ANSWERS:**

| QUES. | ANS. | QUES. | ANS. | QUES. | ANS. | QUES. | ANS. | QUES. | ANS. |
|-------|------|-------|------|-------|------|-------|------|-------|------|
| 1 | C | 11 | A | 21 | B | 31 | B | 41 | A |
| 2 | A | 12 | B | 22 | B | 32 | A | 42 | A |
| 3 | C | 13 | C | 23 | A | 33 | A | 43 | B |
| 4 | A | 14 | A | 24 | B | 34 | B | 44 | D |
| 5 | B | 15 | B | 25 | B | 35 | C | 45 | A |
| 6 | C | 16 | B | 26 | A | 36 | B | 46 | A |
| 7 | A | 17 | B | 27 | B | 37 | D | 47 | B |
| 8 | A | 18 | A | 28 | B | 38 | A | 48 | A |
| 9 | A | 19 | B | 29 | D | 39 | C | 49 | B |
| 10 | C | 20 | B | 30 | A | 40 | A | 50 | A |

# 2. FORCE AND LAWS OF M OTION

## SOME IMPORTANT POINTS

➢ Force is muscular effort which can change the magnitude of velocity the direction of motion the shape or size of an object.

➢ When the applied force on a object by two sides is equal that is called balanced force.

➢ When unbalanced force acting on an object brings the object in motion.

➢ First law of motion states that an object remains in rest or motion unless an unbalanced force act on it motion.

➢ The tendency of an object to maintain its state is called inertia.

➢ The object which has a larger mass has a larger inertia.

➢ The momentum of an object is described by the product of its mass and velocity
   ($p=mv$)

➢ The SI unit of momentum is ($kgm/s^{-1}$)

➢ Momentum gas both direction and magnitude

➢ The second low of motion stats that the rate of change of momentum is dire city proportional to force applied on a object.

➢ Newton's second low of motion force acting on am object.
   ($f=ma$)

➢ Newton's and third low of motion states that every action has a equal and opposite reaction

➢ Action and reacting acts on two different objects.

➢ Forces of action and reaction are equal and opposite

➢ The SI unit of force is newton or $kgms^{-2}$

➢ Law of conservation of momentum shows that sum of momentum before the collision is equal to the sum of momentum after the collision.
   ($m_1u_1+m_2u_2)=(m_1v_1+m_2v_2$)

# 2. FORCE AND LAWS OF MOTION

1.      The SI unit of momentum is?

    a. PaS           b. NS

    c. N/S           d. kgm/S$^2$

2.      Which is the odd one out?

    a. force           b. momentum

    c. mass           d. acceleration

3.      Force is defined as:

    a. mass × velocity        b.  Mass  × acceleration

    c. mass/volume        d. pressure/area

4.      When a force of 35 N acts on a body of mass 7 kg, the body will be accelerated at:

    a. 5 m/s$^2$           b. none

    c. 35 m/s$^2$           d. 245 m/s$^2$

5.      A mass of 7 kg moves at 4m/s. its momentum is

    a. 28NS                b. 70 Ns

    c. 28 Kgm/s$^2$            d. 70 Kgm/s$^2$

6.      When a force of 30 N acts on a body of mass 10 kg, the body will be accelerated at

    a. 30  m/s$^2$           b. 3 m/s$^2$

    c. 300 m/s$^2$           d. 236 m/s$^2$

7.      Which is the odd one?

a. speed                           b. density

c. momentum                        d. mass

8.    The SI unit of force is

a. Pascal                          b. joule

c. Newton                          d. dyne

9.    A mass of 9 kg moves at 2 m/s. its momentum is:

a. 18 kgm/s$^2$                    b. 90 kg m/s$^2$

c. 18 Ns                           d. 90 Ns

10.   An object is moving in a circular path. The force acting on the objet towards the centre of the circle is called the

a. magnetic force                  b. centripetal force

c. gravitational force             d. none of these

11.   Which of the following statement is not true?

a. momentum = mv                   b. f = ma

c. velocity = speed/time           d. none

12.   A body continues in its state of rest or uniform motion in a straight line unless acted upon by a force is a statement of

a. Newton's first law of motion    b. Newton's second law of motion

c. Newton's third law of motion    d. law of gravitation

13.   Action and reaction always act on different bodies in opposite directions according to which laws:

a. First law of motion             b. second law of motion

c. Third law of motion             d. none

14. Who wrote the book named "the little balance":

   a. Galileo galilei          b. M.K Gandhi

   c. both a and b             d . None

15. Galileo galilei was born on

   a. 15th Feb. 1564          b. 16th march 1697

   c. 15th Feb. 1565          d. 15th Feb. 1678

16. An object to resist a change in its state motion or state of rest is

   a. inertia                 b. first law of motion

   c. Second law of motion    d. momentum

17. What is the Si unit of mass and acceleration respectively?

   a. m and kg                b. kg and m/s^2

   c. m/s and m               d. Ns and N/s^2

18. What is the symbol of normal force?

   a. N                       b. Ns

   c. m/s                     d. m/s^2

19. Ship floats on the principal of which force

   a. buoyant force           b. gravitational force

   c. both a and b            d.  None of these

20. Define third law of motion

   a. to every action there is an equal and opposite reaction

   b. if it is at rest it tends to remain at rest

   c. both a and b

d. none

21. a milk tanker filled up to ¾ of its height is moving with a uniform speed on sudden application of the brake, the milk in the tank would

a. move backward                 b. move forward

c. be unaffected                  d. rise upward

22. a mass of 6 kg moves at 2m/s. its momentum is

a. 60Ns                           b. 12Ns

c. 60 kg m/s^2                    d. 12kg m/s^2

23. The resistance offered by an object to an applied force is referred to as

a. inertia                        b. potential

c. reaction                       d. friction

24. Which of the following has more inertia?

a. stone                          b. train

c. five rupees coin               d. pen

25. Which is required to keep a moving object in motion?

a. displacement                   b. force

c. gravitational pull             d. none of these

26. If the set of force acting on an object are balanced, then object must be:

a. at rest                        b. moving

c. accelerating                   d. none

27. Nuclear force is an example of which force

a. contact force                  b. non contact force

c. gravitational force          d. none

28. What is the momentum of an object of mass m, moving with a velocity v?

    a. $(mv)^2$                    b. mv^2

    c. .5 mv^2                  d.mv

29. A spring scale reads 20 N as it pulls a 4 kg object across a table. What is the magnitude of the force exerted by object on the spring scale?

    a. 40N                  b. 20N

    c. 4N                   d. 5N

30. When a bus starts suddenly the bus passengers standing in the bus tend to fall backwards. This is due to:

    a. inertia                b. inertia of motion

    c. inertia of direction      d. none

31. The vector sum of all balanced forces is:

    a. zero                 b. one

    c. Two                 d. three

32. Which law states that the external force is required to change the inertia of the object?

    a. Newton's first law       b. second law of motion

    c. Third law           d. none

33. Which is the example of contact force?

    a. buoyant force        b. friction force

    c. weak nuclear force    d. both a and b

34. 1 N is equivalent to what?

a. 0.5 dyne            b. $10^6$ dyne

c. $10^5$ dyne          d. $10^7$ dyne

35.    At bats man hits a cricket ball which then rolls on a level ground. After
       covering short distance, the ball comes to rest. The ball slow down and stop
       because:

       a. the bats man didn't hit the ball hard

       b. there is a force on the ball opposing the motion

       c. velocity is proportional to the force exerted on the ball

       d. all of these

36.    An object of mass 5kg is moving with a velocity of 4m/s. a constant force of
       20 N acts on the object from opposite direction. What will be the velocity of
       the object?

       a. 12 m/s            b. 0 m/s

       c. 15 m/s            d.16 m/s

37.    Define impulse:

       a. the effect of force applied for a short duration

       b. the product of force and the time duration for which the force is applied

       c. both

       d. none


38.    What is the SI unit of the impulse?

       a. m/s            b. same as of momentum

       c. m/s^2          d. none

39. Which is a type of collision?

    a. elastic collision          b. plastic collision

    c. impulse          d. all of these

40. The interaction between two or more bodies is called:

    a. momentum          b. collision

    c. mass          d. none

41. if an object of mass 9 kg starts from rest and attains a velocity of 18 m/s after 6s, then the force acting on it is:

    a. 27 N          b.108 N

    c. 3N          d. 54N

42. The rate of change of .............is equivalent to force applied:

    a. velocity          b. momentum

    c. displacement          d. density

43. When an object is at rest on a surface, what can you say about the forces on it?

    a. they are unbalanced forces

    b. there is no any forces

    c. all the forces cancel out each other

    d. all the forces are in the same direction

44. Electromagnetic force, gravitational force, strong nuclear force, weak nuclear force:

    For the above forces: which force is strongest?

    a. electromagnetic          b. gravitational

c. strong nuclear    d. weak nuclear

45. for the given forces: which force acts downwards on the surface?

 a. weight      b. air resistance

 c. friction      d. none

46. .............. forces will have no effect on the momentum of an object:

 a. unbalanced force   b. balanced force

 c. opposite forces    d. none

47. How are force, mass and acceleration related?

 a. F=m/a     b. F=a/m

 c. F=ma     d. m=Fa

48. When an object reaches its maximum velocity when falling through a fluid, what do we call it?

 a. acceleration    b. deceleration

 c. constant velocity   d. terminal velocity

49. The direction of friction is always ..........to the direction of the object motion:

 a. equal      b. opposite

 c. unrelated     d. related

50. When an object is moving fast through a fluid how does this affect the force of friction on it?

 a. the forces of friction is greater  b. the forces of friction is smaller

 c. the forces of friction unaffected  d. the forces of friction is same

51. A car of mass 1000kg can produce an acceleration of 8m/s^2. Calculate the force produced by the engine ignoring friction:

a. 10000 N          b. 8000 N

c. 125 N          d. 80000 N

**ANSWERS:**

| QUES. | ANS. | QUES. | ANS. | QUES. | ANS. | QUES. | ANS. | QUES. | ANS. |
|-------|------|-------|------|-------|------|-------|------|-------|------|
| 1 | C | 11 | C | 21 | B | 31 | A | 41 | A |
| 2 | C | 12 | A | 22 | B | 32 | A | 42 | B |
| 3 | B | 13 | C | 23 | A | 33 | D | 43 | C |
| 4 | A | 14 | A | 24 | B | 34 | C | 44 | A |
| 5 | A | 15 | A | 25 | B | 35 | B | 45 | A |
| 6 | B | 16 | A | 26 | A | 36 | B | 46 | B |
| 7 | C | 17 | B | 27 | B | 37 | C | 47 | C |
| 8 | C | 18 | A | 28 | D | 38 | B | 48 | D |
| 9 | C | 19 | A | 29 | B | 39 | A | 49 | B |
| 10 | B | 20 | A | 30 | A | 40 | B | 50 D | |
| | | | | | | | | 51 B | |

# 3. GRAVITATION

## SOME IMPORTANT POINTS

➢ The body moving along the circular path is acting towards the centre the centre by a force this force is centripetal force.

➢ The force of attraction between a object to any other object in universe this force is known as gravitational force.

➢ Every object in universe attract every other object in universe.

➢ Low of gravitation the force of attraction between any two object in universe is proportional to product of their masses and inversely proportional to the square of distance between them. The force is along live joining their centre.

➢ $F = G \, M*m/d^2$ [force, M, m=mass of two object d=distance]
G=universal gravitation constant.
The S.I unit of G is $Nm^2kg^{-2}$ and value of $G = 6.673*10^{-11} \, Nm^2kg^{-2}$

➢ Gravity is a force of gravitation of earth.

➢ When an object falls towards the earth under gravity its called free fall

➢ G is the acceleration due to gravity the unit of g is $m/s^2$ and value of $g = 9.8 m/s^2$ $g = G \, M/d^2$ [where M=mass of earth d=distance between object & earth]

➢ The value become grater at the poles than at equator

➢ Mass of an object is measure of its inertia mass of an object is constant and does not change place to place.

➢ It is scalar quantity and its unit is kilogram (kg)

➢ The weight of an object is a force with which it is attracted to wards the earth

➢ Weight =mass*gravity
F=mg, unit of weight is Newton

➢ it is a vector quantity

➢ Weight of an object on moon is 1/6th the weight of that object on earth.

➢ The perpendicular force acting on an object is known as thrust.

➢ The unit of thrust is Newton

- The thrust per unit area is known as pressure
- The S.I unit of pressure is $N/m^2$ bat in hours of scientist base Pascal the S.I unit of pressure called Pascal denoted by pa
- The force acting on smalls area exerts a larger pressure
  This is the reason of why a nail has pointed tip
- All liquids and gases are fluids
- The upward force exerted by a liquid on object is known as up thrust or buoyant force
- Archimedes principle when a body is immersed fully or partially in a fluid. It experiences an upward force that is equal the weight of fluid displaced by it
- Archimedes principle used in designing ships submarines, Lactometers, hydrometer etc.
- The relative density of a substance is the ratio of its density to that of water
  Relative density= density substances /density of water

# 3. GRAVITATION

1.  Who calculated the value of G?

    a. Newton    b. Cavendish    c. Galileo    d. Chadwick

2.  The value of G:

    a. $9.8 \text{ m/s}^2$    b. $60673 * 10^{-11} \text{ Nm/kg}^2$

    c. $9.998 * 10^{-12} \text{ Nm}^2 /\text{kg}^2$    d. $9.998 * 10^{-11} \text{ m/s}^2$

3.  Where the radius of earth is more:

    a. pole    b. equator

    c. both a and b    d. everywhere same

4.  The value of g is:

    a. $6.023 * 10^{23}$    b. $6.673 * 10^{-11}$

    c. $6.673 \text{ m/s}^2$    d. $9.8 \text{ m/s}^2$

5.  The mass of a man is 50 kg on the earth then what will be the value of the mass of that man on moon:

    a. 7kg    b. 50/6kg

    c. 50/2kg    d. 50kg

6.  SI unit of weight is:

    a. m    b. kg

    c. N    d. ton

7.  Weight is a ……………………. quantity:

    a. scalar    b. vector

    c. both a and b    d. Constant

8.  What is the radius of the earth?

    a. $5.98 * 10^{24}$              b. $1.74 * 10^6$

    c. $1.74 * 10^{24}$              d. $6.37 * 10^6$

9.  What is the mass of the moon?

    a. $1.74 * 10^{29}$              b. $7.36 * 10^{22}$

    c. $7.36 * 10^{29}$              d. $7.36 * 10^{59}$

10. Mass of an object is 24kg. What is the weight on earth?

    a. 24/6N                         b. 4N

    c. 235.2 kg                      d. 235.2 N

11. force exerted perpendicular to the surface is known as:

    a. pressure                      b. thrust

    c. burst                         d. buoyant force

12. The unit of pressure is:

    a. $kg/s^2$              b. $N/m^2$

    c. $N/kg^2$              d. $Nm^2/kg^2$

13. The SI unit of thrust is:

    a. kg                            b. N

    c. Pa                            d. kgm/s

14. The SI unit of pressure is:

    a. N                             b. Pa

    c. P                             d. kgm/s

15. The thrust of a unit area is known as:

a. power                    b. buoyant force

c. Pascal                   d. pressure

16.   What is the Si unit of density?

a. $kg/cm^3$                b. $kg/m$

c. $kg/m^2$                 d. $kg/m^3$

17.   Magnitude of the buoyant force depends on..........................of the fluid displaced:

a. quantity                 b. volume

c. density                  d. height of the container

18.   The upward force exerted by the fluids on the object is known as:

a. thrust                   b. energy

c. up thrust                d. power

19.   Lactometers and hydrometers are based on which principle's:

a. Newton's                 b. Henry Cavendish

c. J.J Thomson's            d. Archimedes

20.   Lactometer is used to measure:

a. purity of water          b. purity of acid and base

c. purity of milk sample    d. all of the above

21.   What is the density of gold (in kg/m^3):

a. 19037                    b. 19012

c. 19300                    d. 1880

22. The relative density of a material is 20.8. The density of water is 1000 kg/m$^3$. Then what is the density of that material (in kg/m$^3$):

   a. 2080              b. 1800

   c. 20800             d. 208000

23. When we throw an object then it falls:

   a. in any direction      b. upward

   c. downward          d. both b and c

24. The motion of moon around earth is due to ...................force:

   a. gravitational       b. nuclear

   c. electromagnetic     d. muscular

25. According to the universal law of gravitation the force between object is directly proportional to the product of their:

   a. speed             b. distance

   c. mass              d. weight

26. The universal law of gravitation is applicable to .........................:

   a. small bodies       b. big bodies

   c. all of these       d. terrestrial bodies

27. Which force binds us to earth?

   a. Muscular force     b. gravitational force

   c. nuclear force      d. centripetal acceleration

28. When an object is in the free fall it's.........of the motion do not change:

   a. speed             b. velocity

c. direction   d. displacement

29. When an object is in the free fall its velocity changes due to:

  a. acceleration   b. gravitational force

  c. speed     d. direction

30. When an object is in the free fall its velocity:

  a. remains same   b. Increases

  c. decreases    d. depends on the nature of the object

31. Due to which force all objects in the universe attract each other:

  a. electromagnetic force   b. nuclear force

  c. gravitational force    d. centripetal force

32. The weight of a man on the moon is 6 N. What is its mass on the moon? Given $g=10m/s^2$:

  a. 100/ (98*36) kg    b.1/98 kg

  c. 36/98 kg      d. 3.6kg

33. The mass of an object on the moon is 10kg what is its weight on earth:

  a. 0.00098N     b. 0.98 N

  c. 9.8 * 10^-2     d. 9.8* 10^-3N

34. The density of silver is (in kg/m^3):

  a. 1000      b. 10080

  c.10800      d.11870

35. An object of mass 20 kg falls from a certain height. Calculate the force it experience:

a. 20N                               b. 199N

c. 196N                              d. 197N

36.     An object fall from a certain height it experiences a force of 0.966N. Then
        calculate its weight on moon

        a. 161N                          b.966N

        c. 1.61 N                        d.966 kg

37.     An object falls from a certain height. It experiences a force of 0.966 N
        forces then what will be its weight on moon?

        a.   $161*10^{-3}$ N             b. 0.966N

        c.   $161*10^{-2}$ N             d. 0.966Kg

38.     The mass of an object remains same on…….?

        a. only earth                    b. only moon

        c. only planets                  d. all over the universe

39.     The value of acceleration due to gravity is?

        a. same at poles and equator     b. more at equator

        c. more at pole                  d. remains same every where

40.     he atmosphere is held to the earth by?

        a. earth magnetic field     b. air

        c. atmosphere               d. gravity

41.     When the distance between two mass is doubled then the force becomes?

        a. 1/6 times                b. 4 times

        c. ½ times                  d. ¼ times

42. The relationship between "G" and "g" is?

    a. $G = gM/(d^2)$                 b. $g = GM/(d^2)$

    c. $g = Md^2/g$                    d. $g = d^2/GM$

43. Newton's Law of gravitation is applicable on?

    a. earth                  b. moon

    c. all over the universe          d. atmosphere

44. The density of water is?

    a. $1100 \ kg/m^3$                b. $1000 \ kg/m^4$

    c. $1000 \ kg/m^3$                d. $1001 \ kg/m^3$

45. The mass of an object on the moon is 1000 kg. What is its mass on the earth?

    a. 1000kg                 b. 1000/6 kg

    c. 6000 kg                d. 166.666 kg

46. When the masses of the two objects are doubled. The gravitational force between them becomes?

    a. ¼ times                b. 4 times

    c. 1/6 time               d. 6 times

47. "G" is?

    a. gravitational constant         b. power

    c. energy                 d. gravity

48. What is the unit of "G"?

    a. $Nm^2/kg^2$                    b. $Nm^2kg^2$

c. Nm/kg$^2$                    d. N/kg$^2$

49.     What is the value of "G" on moon?

a. 6.673*10$^{-23}$                    b. 6.673*10$^{-11}$

c. 6.673*10$^{-12}$                    d. 6.689*10$^{-13}$

50.     Mass is?

a. scalar quantity          b. vector quantity

c. none of these            d. vector having no direction

**ANSWERS**

| Ques. | Ans. | Ques. | Ans. | Ques. | Ans. | Ques. | Ans. | Ques. | Ans. |
|-------|------|-------|------|-------|------|-------|------|-------|------|
| 1 | B | 14 | B | 27 | B | 40 | D | | |
| 2 | B | 15 | D | 28 | C | 41 | D | | |
| 3 | A | 16 | D | 29 | B | 42 | B | | |
| 4 | D | 17 | C | 30 | B | 43 | C | | |
| 5 | D | 18 | C | 31 | C | 44 | C | | |
| 6 | C | 19 | D | 32 | D | 45 | C | | |
| 7 | B | 20 | C | 33 | C | 46 | B | | |
| 8 | D | 21 | C | 34 | B | 47 | A | | |
| 9 | B | 22 | C | 35 | B | 48 | A | | |
| 10 | D | 23 | C | 36 | B | 49 | B | | |
| 11 | B | 24 | A | 37 | A | 50 | A | | |
| 12 | B | 25 | C | 38 | D | | | | |
| 13 | B | 26 | C | 39 | C | | | | |

# 4. WORK AND ENERGY

## SOME IMPORTANT POINTS

➤ When some force applied on a object and object must be displaced when we say that work is done.

➤ The unit of work is Joule.

➤ When 1N force applied on a object and object displaced 1m then we say that work done is 1J.

➤ 1J=1N*1m.

➤ The capacity to can do a work is known as energy. The unit of energy is joule also.

➤ There are many types of energy like kinetic energy, potential energy, mechanical energy, solar energy, nuclear energy etc.

➤ Object posses energy due to motion it said to be energy.

➤ Increased in speed kinetic energy is also increased.

➤ Objects possess energy due to change its height, shape, size it said to be potential energy.

➤ When work done against the gravity the energy possessed gravitational potential energy.

➤ Increase in height gravitational potential energy.

➤ Law of conservation of energy states the energy gets transformed one form to another; it can neither be created nor be destroyed.

➤  The total energy before the transformation and after transformation remains same it is valid in all situation.

➤ Potential energy+ Kinetic energy=Constant.

➤ The rate of doing work is known is power.

➤ The unit of power is watt and 1watt=1J/1sec.

➤ 1KW=1000W

➤ 1 horse power=746W.

➤ The energy used in 1hour of rate of 1KW known as1KWh.

➤ $1KWh=3.6*10^6$ Joule.

➤ Work=Force*displacement

- W=F*S
- Kinetic energy=1/2*mass*(velocity)$^2$.
- $E_k=1/2mv^2$
- $E_k=P^2/2m$                         (P=momentum, m=mass)
- Potential energy=mass*velocity*acceleration due to gravity
- $E_p$=m*g*h=mgh
- Power=Work/time
- 1unit=1KWh.

# 4. WORK AND ENERGY

1. What two conditions need to be satisfied for work to be done?

   a. a force should act on object and the object must be displaced.

   b. a force should act on object and the object should not be displaced.

   c. a force should act on object and the shape of the object must be changed.

   d. both a & c.

2. What is the symbol of the work done?

   a. F          b. N          c. W          d. G

3. What is not the formula of the work done?

   a. F.v          b. F.s          c. ½ mv²          d. mgh

4. What is the SI unit of work?

   a. watt          b. Nm          c. N          d. N/s

5. A force of 7N is acting on an object and it is displaced through 20cm in direction of the force. Then the work done is?

   a. 20j          b. 24j          c. 19j          d. 21j

6. A force of 9N is acting on an object and it is displaced through 3m in direction of the force. Then the work done is?

   a. 18j          b. 7.2j          c. 1.8j          d. 1.6j

7. Work done by a force can be?

   a. negative          b. positive          c. both a and b          d. none of these

8. A man lifts a stone of 3kg from the ground and put it on his head 1.3m above the ground. Determine the work done by him on the stone.

a. 39j      b. 40j      c. 41j      d. 37j

9.      Which is the biggest source of energy for us?

a. earth      b. moon      c. sun      d. none of these

10.      What is the SI unit of energy?

a. Joule      b. Newton      c. Watt      d. Kilogram

11.      1 Kilo Joule (Kj) =?

a. 1000j      b. 746j      c. 900j      d. 786j

12.      Kinetic energy + potential energy = ............... energy.

a. Mechanical      b. nuclear      c. electrical      d. chemical

13.      Object in motion posses which type of energy?

a. potential energy      b. light energy

c. chemical energy      d.kinetic energy

14.      When kinetic energy increases?

a. when potential energy increases      b. when potential energy decreases

c. when mechanical energy decreases      c. none of these

15.      What is the formula of kinetic energy?

a. ma      b. $1/2mv^2$      c. mgh      d. F.s

16.      An object of mass 32kg is moving with a uniform velocity of 5m/s. what is the kinetic energy possessed by the object.

a. 160j      b. 1300j      c. 400j      d. 650j

17.      a object of mass 10kg moving with a uniform velocity of 2m/s. what is the kinetic energy possessed by the object?

a. 20j          b. 30j          c. 40j               d.10j

18.     An object of mass 20kg moving with uniform velocity of 3m/s. what is the kinetic energy possessed by the object?

a. 126j          b. 546j          c. 710j               d. none of these

19.     What is the work to be done to increase the velocity of a bus from 18km/h to 35km/h if the mass of the bus is 4500kg?

a. 168640j          b. 168750j          c. 167750j               d. none of these

20.     What is the work to be done to increase the velocity of car from 9km/h to 54km/h if mass of car is 2500kg?

a. 228750j          b. 218650j          c. 218750j               d. none of these

21.     What is the work to be done to increase the velocity of a car from 20km/h to 60 km/h if mass of the car is 1800kg?

a. 22222.22j          b. 222222.22j

c. 2222222.22j          d. none of these

22.     What is the formula of potential energy?

a. F.s          b. mgh          c. $1/2mv^2$          d. w/t

23.     Find the energy possessed by an object of mass 20kg when it is at a height of 8m above the ground. Given g = 9.8 ms$^{-2}$.

a. 1578j          b. 1567j          c. 1558j          d. 1568j

24.     Find the energy possessed by an object of mass 100kgif it is at height of 3m above the ground. Given g = 9.8 ms$^{-2}$.

a. 2740j          b. 2940j          c. 2840j          d. 2640j

25. An object of mass 20 kg is at a certain height above the ground if the potential energy of the object is 1960j. Find the height at which the object is with respect to ground.

    a. 8m          b. 12m          c. 10m          d. 11m

26. An object of mass 15 kg is at a certain height above the ground if the potential energy of the object is 300j. Find the height at which the object is with respect to ground.

    a. 2m          b. 3m          c. 7m          d. 4m

27. When an object falls on ground from certain height "h". Its potential energy will be changes into ...............energy.

    a. chemical          b. electrical          c. mechanical          d. kinetic

28. Rate of doing work is known as?

    a. power                    b. kinetic energy

    c. acceleration          d. none of these

29. What is the SI unit of power/

    a. Joule          b. Newton          c. Watt          d. None of these

30. 1 Watt = ....?

a.    1 Joule/s          b. 1 N/s          c. 1 Hz/s          d. 1 m/s

31. 1 Horse Power = ...............watt?

    a. 756          b. 786          c. 746          d. None of these

32. A girl having weight 200 N climb up a rope through a height of 4 m. Girl takes 10s to accomplish this task. What is the power of this task?

    a. 800W          b. 80W          c. 50W          d. 70W

33. A boy of weight 150N runs up a staircase of 60 steps in 10s. if the height of each step is 20 cm. find the power?

    a. 180W        b. 1800W        c. 280W        d. 780W

34. 1 KWh =..............J.

    a. $3.6*10^5$        b. $36*10^3$        c. 360000        d. None of these

35. An electric fan f 150 watt is used for 5h per day. Calculate the "units" of energy consumed in one day by the fan.

    a. 75        b. 0.75        c. 8.5        d. 0.7

36. An electric bulb of 20W and two fan of 15W and television of 250W are used for 7h per day. Calculate the total units of energy consumed in 7 days by these appliances.

    a. 20.78        b. 20.67        c. 20.14        d. None of these

37. Work done on an object by a force would be zero if the displacement of the object is..........................?

    a. zero                b. more than zero

    c. work can't be zero        d. None of these

38. Which has the same unit as of work?

    a. power        b. force        c. energy        d. None of these

39. Which can neither be created nor be destroyed?

    a. energy        b. mass        c. both a and b        d. none of these

40. The energy used in 1 hour at the rate of 1KW is called....................?

    a. 1kwh        b. 1kw        c. 1 h        d. none of these

41. The potential energy of a free falling object decreases and kinetic energy........?

a. also decreases    b. increases    c. remains same    d. none of these

42. A body is under action of the equal and opposite forces. The work done by the body is?

    a. 49J    b. 0    c. -49J    d. 35J

43. If the mass of the body becomes 4 times, its kinetic energy?

    a. increases 4 times    b. gets doubled

    c. remains same    d. increases 8 times

44. If the speed of the train is increased by 4 times. Its kinetic energy will be increased b y?

    a. 4 times    b. 12 times    c. 20 times    d. 16 times

45. Watt is the unit of?

    a. rate of doing work    b. rate of doing power

    c. rate of change of velocity    d. None of these

46. Two bodies of equal mass are kept at height of h and 2.5h, respectively. The ratio of their potential energy is?

    a. 5:4    b. 5:2    c. 4:2    d. 2:5

47. $kgm^2s^{-3}$ is associated with?

    a. energy    b. momentum    c. force    d. power

48. 1MJ = ..........?

    a. $10^6J$    b. $10^4J$    c. $10^5J$    d. $10^3J$

49. A body is thrown up with kinetic energy of 25J, if it attains maximum height 10m. Find the mass of the body. Given g = $10m/s^2$.

    a. 0.2 kg    b. 0.7 kg    c. 0.15kg    d. 0.25kg

50.     How much time does it take to perform 250j of work at a rate of 25W?

        a. 25s          b. 20s          c. 50s          d. 10s

Answers:

| Que. | Ans. | Que. | Ans. | Que. | Ans. | Que. | Ans. | Que. | Ans. |
|------|------|------|------|------|------|------|------|------|------|
| 1 | D | 11 | A | 21 | B | 31 | C | 41 | B |
| 2 | C | 12 | A | 22 | B | 32 | B | 42 | B |
| 3 | A | 13 | D | 23 | D | 33 | A | 43 | A |
| 4 | B | 14 | B | 24 | B | 34 | D | 44 | D |
| 5 | D | 15 | B | 25 | C | 35 | B | 45 | A |
| 6 | C | 16 | C | 26 | A | 36 | D | 46 | D |
| 7 | C | 17 | A | 27 | D | 37 | A | 47 | D |
| 8 | A | 18 | A | 28 | A | 38 | C | 48 | A |
| 9 | C | 19 | B | 29 | C | 39 | C | 49 | D |
| 10 | A | 20 | C | 30 | A | 40 | A | 50 | D |

# SOUND

## SOME IMPORTANT POINTS

➤ Sound produced by vibration of object.

➤ Sound is the form of mechanical energy.

➤ Propagation of sound waves as series of compression and rarefaction called longitudinal wave.

➤ Change in density from maximum value to minimum value.

➤ The number of complete oscillation in density per second of a sound wave called frequency.

➤ Frequency is denoted by $\mu$.

➤ S.I unit of frequency is hertz (Hz).

➤ Distance between two consecutive compressions and two consecutive rarefactions called wavelength. It is denoted by $\lambda$ (lemda).

➤ Speed of sound V = distance/time, $\lambda/t = v = \lambda/t$.

➤ Frequency is directional proportional to pitchnes.

➤ Amplitude is directly proportional to loudness.

➤ Speed of sound in solid > Speed of sound in liquid >Speed of sound in gases.

➤ Speed of sound increases with increasing in temperature.

➤ Reflection of sound: Angle of reflection is equal to angle of incidence $\rightarrow \angle r = \angle i$ reflected wave, incident wave and normal all lies in same plane.

➤ Repetition of sound by reflection from an absolute called echo.

➤ Human ears can hear only 20Hz to 20 KHz of frequency.

➤ The longitudinal wave's frequencies below 20Hz or called infrasonic, longitudinal waves frequencies lie 20 KHz called ultrasonic.

➤ SONAR is sound navigation and ranging used to measure distance in sea.

# 5. SOUND

1. How does sound is produced?

   a. mass     b. vibration  c. weight     d. energy

2. Sound is a form of?

   a. energy     b. motion     c. power     d. force

3. For transferring sound energy we need:

   a. propagation     b. vibration  c. medium   d. all of these

4. The sound of human voice is produced due to vibration of:

   a. Brain     b. Neck     c. Vocal chords     d. teeth

5. Sound can't travel through:

   a. solid     b. liquid     c. gas     d. vacuum

6. A particle of the medium in contact with the vibrating object is first displaced from its:

   a. equilibrium position     b. Rest position

   c. Far position     d. all of these

7. After displacing the adjacent particle the first particle comes back to its original position in a longitudinal wave:

   a. True     b. false     c. depends on temperature     d. None of these

8. Sound waves are characterized by the motion of particle in the medium and it is called:

   a. Mechanical wave     b. Acoustic wave

   c. Electromagnetic wave     d. None of these

9.    In which matter sound travels fast?

a. liquid     b. gas     c. jelly     d. solid

10.   When vibrating objects move forward, it pushes the air in front of it creating a region of high pressure, which is called?

a. compression     b. rarefaction     c. pushing   d.  all of these

11.   Can we hear sound in vacuum?

a. yes     b. No     c. sometimes     d. depends on speed of sound

12.   Sound waves are:

a. longitudinal     b. acoustic   c. transverse     d.  None of these

13.   In the wave the individual particle of medium moves in a direction parallel to the direction of motion is known as:

a. Propagation     b. Wave front     c. Vibrating object   d. all of these

14.   What is the SI unit of Wavelength?

a. meter     b. ampere   c. Newton   d. joule

15.   In a transverse wave particle oscillate along the line of:

a. wave propagation                    b. Wave front

c. perpendicular to the motion          d. none of these

16.   In SONAR we use?

a. ultrasonic wave                    b. infrasonic wave

c. radio wave                         d. audible sound

17.   Sound travels fast in which season?

a. summer                    b. winter

c. rainy                           d. spring

18.  When we change a feeble sound to loud sound we increase its?

    a. frequency                    b. amplitude

    c. velocity                     d. wavelength

19.  Melody of sound depends on:

    a. frequency                    b. amplitude

    c. velocity                     d. wavelength

20.  Earthquake produces which type of sound before the main shock wave begins:

    a. ultrasound                   b. infrasound

    c. audible sound                d. none of these

21.  Infrasound can be heard by:

    a. dog            b. bat

    c. rat            d. rhinoceros

22.  Symbol of wavelength is:

    a. $\sigma$        b. $\lambda$        c. v        d. $\pi$

23.  The frequency of the sound wave is represented by:

    a. v        b. $\lambda$        c. $\sigma$        d. $\pi$

24.  The wave propagation vibrates in a direction perpendicular to the direction of propagation of wave:

    a. transverse wave motion.       b. longitudinal wave.

    c. simple wave                   d. none of these

25. Speed of sound is denoted by:

   a. V          b. λ          c. σ          d. π

26. Speed of sound is equal to :

   a. (distance/time taken) = λ/T          b.  P = MV

   c.  P = F/A                             d. None of these

27. Choose the characteristics of sound waves:

   a.  frequency          b. pressure

   c.  explosion          d.  rarefaction

28. Sound pollution is measured in:

   a. decibels            b. amplitude

   c. velocity            d. wavelength

29. Wave has frequency of 50Hz what is its time period in second?

   a. 0.02          b. 2          c.  0.0005          d. none of these

30. The loudness or softness is determined basically of its:

   a. frequency          b. amplitude          c. Velocity          d. none of these

31. Full form of SONAR is:

   a.  sound navigation and rarefaction

   b. sound navigation and rairing

   c.  sound navigation and ranging

   d. none of these

32. Note is a sound of:

   a. mixture of several frequencies

b. mixture of two frequencies only

c. a single frequency

d. always unpleasant to listen

33. Sound travels in air if:

a. particle of medium travel from one place to another

b. there is no moisture in the atmosphere

c. disturbance

d. all of these

34. When we change feeble sound to loud sound, we increases its:

a. frequency      b. amplitude      c. velocity      d. wavelength

35. The half wavelength is:

a. $\lambda/2$      b. $2\lambda$      c. $2.5\lambda$      d. $\lambda^2$

36. A SONAR device on submarine sends out a signal and receives an echo 5s later. Calculate the speed (in m/s) of sound in water if the distance of the object from the submarine is 3625m?

a. 1456      b. 1450      c. 2036      d. 1540

37. Ultrasound is used in:

a. cleaning      b. in medical

c. both a and b      d. none of these

38. SI unit of frequency is:

a. joule      b. Newton      c. Pascal      d. Hz

39. The frequency of sound is 100Hz. How many times does it vibrate in a minute?

a. 600                    b. 6000                    c. 60                    d. 60000

40.    Speed of wave is :

       a. F/A          b. $\lambda$/T          c. $mc^2$                    d.  none of these

41.    A sound of single frequency is:

       a. tone                    b. note          c. Infrasonic sound          d. none of these

42.    The sensation of sound persists in our brain for about:

       a. 0.2s                    b. 0.1s          c. 1s          d. 0.02s

43.    Echo is a:

       a. once reflection of sound                    b. multiple reflection of sound

       c. multiple reflection                    d. none of these

44.    The repeated reflection that results in their persistence of sound is called:

       a. Refraction                    b. reverberation

       c.  both a and b                    d. none of these

45.    Uses of multiple reflection of sound is:

       a. cleaning                    b. stethoscope          c.  microphones          d. both c and d

46.    The number of complete oscillation per unit time is called:

       a. amplitude          b. ultrassound          c.  frequency          d. none of these

47.    The outer ear is called:

       a. cochlea                    b. pinna          c.  anvil                    d. stirrup

48.    Bat produces:

       a. infrasonic sound                    b. ultrasonic sound

       c. beam of sound                    d.  None of these

49. The amount of sound energy passing each second through unit area is called:

a. density of sound

b. amplitude

c. intensity of sound

d. all of these

50. Ultrasound may be employed to break the small:

a. stones
b. light particle
c. atoms
d. both a and c

ANSWERS:

| QUE. | ANS. | QUE. | ANS. | QUE. | ANS. | QUE. | ANS. | QUE. | ANS. |
|------|------|------|------|------|------|------|------|------|------|
| 1 | B | 11 | B | 21 | D | 31 | C | 41 | B |
| 2 | A | 12 | A | 22 | B | 32 | C | 42 | B |
| 3 | C | 13 | A | 23 | A | 33 | C | 43 | C |
| 4 | C | 14 | A | 24 | A | 34 | B | 44 | B |
| 5 | D | 15 | C | 25 | A | 35 | A | 45 | D |
| 6 | A | 16 | A | 26 | A | 36 | B | 46 | C |
| 7 | A | 17 | B | 27 | D | 37 | C | 47 | B |
| 8 | A | 18 | B | 28 | A | 38 | D | 48 | B |
| 9 | D | 19 | A | 29 | A | 39 | A | 49 | C |
| 10 | A | 20 | B | 30 | B | 40 | B | 50 | A |

# 6. LIGHT

## SOME IMPORTANT POINTS

- LAWS OF REFLECTION:
    1. Angle of incidence is equal to angle of reflection
    2. The incident ray, reflection rays the normal at the point of incident all lie in same plane.
- Concave mirror are converging mirror convex mirror are diverging mirror.
- The center of reflecting surface of spherical mirror is pole.
- A imaginary line passing through the pole and centre of curvature of a spherical mirror is called principal axis.
- The distance between the pole and principal focus of the spherical mirror is called focal length it is denoted by f.
- The relationship between the radius of curvature and focal length spherical mirror is shows by formula $F=R/2$.
- In convex mirror image is formed virtual and diminished.
- In concave object placed at between P and F image is virtual enlarged.
- Concave mirror ate used in headlight of vehicles in field of solar energy in barber shop etc.
- Convex mirror are used in rear view of traffic in vehicles and in search mirror in big shops.
- Mirror formula : $1/v+1/u=1/f$
- The ratio height of image and height of object is known as magnification of mirror

$M=n^1/h=-v/u$ [for mirror]

- The change in direction of light when it passes from one transparent medium to another is called refraction of light.
- Laws of refraction: The incident ray refracted ray and normal to surface of separation of two transparent media at point of incidence all lie in some plane.

➢ The ratio of sine of angle of incidence to the sins of angle of refraction is constant for a given pair of medium and colour.

➢ The value of sin i/sin r for a ray of right passing through one medium to ant there is known as refraction index.

➢ Spherical lens are types:

➢ Concave lens and convex lens.

➢ Concave lens are diverging lens and convex lens are converging lens.

➢ In concave lens image formed is virtual and diminished.

➢ In convex lens image formed is real, diminished, enlarged and same size.

➢ Lens formula :$1/v-1/u=1/f$

➢ Magnification of lens : $m=h'/h=v/u$

   h'=image height ,h=object height , v=image distance ,u=object distance

➢ The reciprocal of focal length is known power of lens. Its S.I. unit is Dioptre and denoted by D.

➢ Power of concave lens is negative and convex lens has power positive.

# 6. LIGHT

1. The image formed by a plane mirror is always

   (a)    Real and erect               (b) Virtual and inverted

   (c)    Real and inverted  (d)    virtual and erect

2. Which mirror always gives virtual and erect image?

   (a) Concave mirror (b) Convex mirror  (c) Plane mirror     (d) both (b) & (c)

3. The reflecting surface of a mirror curved inward is known as:

   (a) Concave  (b) Convex   (c) Plane     (d) both (a) & (b)

4. The centre of reflecting surface of a spherical mirror is known as:

   (a) Pole               (b) Polo               (c) Centriod           (d) point

5. The centre of Curvature of Concave mirror lies:

   (a) In backward           (b) In front of its reflecting surface

   (c) On the mirror   (d) none of these

6. The relationship between P and f of a spherical mirror is :

   (a) P=F               (b) P=2F     (c) P=2f               (d) p=f/2

7. The image formed by a concave mirror is always

   (a) Real and erect  (b) virtual and inverted

   (c) Real and inverted     (d) None of these

8. Which type of mirror forms a real and same size image of the object:

   (a) Plane     (b) Concave (c) Convex    (d) both (a) & (c)

9.

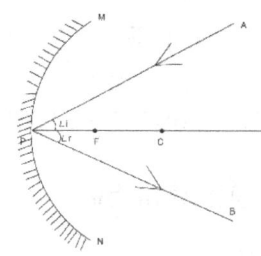

In the figure ∟i = 60⁰, then find ∟BPN=?

(a) 60⁰            (b) 30⁰                    (c) 90⁰                    (d) 45⁰

10.  The focal length of a mirror is 29.5m then find its radius of curvature?

(a) 29.5            (b) 59        (c) 59.5            (d) None of these

11.  Which mirror is used as a rear view mirror in vehicles?

(a) Concave  (b) Plane      (c) Convex    (d) All of the above

12.  The focal length of a mirror is 28cm, object distance is 15cm then find the image distance.

(a) 420/13 m        (b) 420/13 cm        (c) 13/420 cm        (d)  none of these

13.  The magnification of an object is -0.5, then find the nature and size of the image

(a) Real and enlarged              (b) virtual and diminished

(c) Real and Diminished          (d) both (a) & (c)

14.  The height of the image is 0.15, and then finds the nature of the image

(a) Real and inverted      (b) virtual and erect

(c) Real and erect          (d) virtual and inverted

15.  The object distance is always:

(a) Positive  (b) negative (c) depend on the image (d) Depend on the mirror

16.  The distance between the pole and centre of the mirror is known as:

(a) PC          (b) Radius of curvature          (c) CP          (d) All of above

17. Which of the following mirror formed virtual and enlarged image?

(a) Convex   (b) Concave (c) Plane      (d) All of above

18. The image formed on the focus of a mirror is always:

(a) Real and inverted      (b) Highly Diminished Point size

(c) Virtual and erect      (d) Enlarged

19. The S.I. unit of the focal length is:

(a) Diopter   (b) cm          (c) m  (d)  m$^2$

20. Which mirror is used in the rear view mirrors in vehicles?

(a) Plane      (b) Concave (c) both (a) & (b)   (d) Convex

21. The magnification of a plane mirror is:

(a) -1  (b) 1   (c) $\pm$1 (d) All of the above

22. The image distance and the object distance of a mirror is equal. What type of mirror it is:

(a) Concave  (b) Plane      (c) Convex   (d) Spherical

23. Which mirror always forms virtual diminished image

(a) Plane      (b) Convex   (c) Concave (d) None of these

24. The relationship between m, v and u is:

(a) m=v/u    (b) mu=-v    (c) –m=u/v   (d)  m=uv

25. If the Radius of the mirror is 10cm, object distance = 5cm. Then find the focal length of the mirror.

(a) 2.5cm    (b) 5cm          (c) -2.5cm    (d) 2/5

26. The angle of incident is equal to the angle of ..........when light refracted through a glass slab.

    (a) refraction        (b) emergent

    (c) reflection        (d)  Angle of the prism

27. When light passing through one medium to other, its direction slightly charge. This cause is known as:

    (a) reflection        (b) refrection        (c) scattering        (d)  none of these

28. In Rear medium the speed of the light is:

    (a) slower    (b) faster    (c) same    (d)  normal

29. Which is the snell's laws of refraction:

    (a) <i/<r = constant        (b) <r/<I = constant

    (c) <i/<e= constant (d) <e/<r=constant

30. The speed of the light faster in:

    (a) air        (b) Water    (c) Vacuum  (d)  Solid

31. In which medium light ray bent towards the normal:

    (a) denser    (b) rarer        (c) normal    (d)  both (a) & (c)

32. The speed of the light in vacuum is:

    (a) $3 \times 10^8$ Km/S    (b) $3 \times 10^8$ m/s    (c) $3 \times 10^{-8}$ m/s    (d)  both (a) & (c)

33. Which lens is thicker at the middle as compared to the edges?

    (a) Convex    (b) Concave (c) Plane        (d)  bifocal lens

34. Which lens is bulging outwards?

    (a) Convex    (b) Concave (c) Plane        (d) bifocal lens

35.   A lens has _____ focus point:

(a) 1   (b) 2   (c) 3   (d) 0

36.   The central part of the lens is known as:

(a) Pole                (b) optical centre   (c) radius      (d) centre

37.   Which of the following gives a real as well as virtual image?

(a) Convex mirror, concave lens        (b) Convex lens, Concave lens

(c) Concave mirror, convex lens        (d) Concave lens, Convex lens

38.   If the power of a lens is -4diopter then what type of lens it is:

(a) Convex   (b) Concave (c) bifocal     (d)  both (b) & (c)

39.   The Power of a lens is -1diopter, and then find its radius

(a) -1/2                (b) 1   (c) 2   (d) -2

40.   The focal length of a Concave lens is 100cm. the find it power

(a) 100 dioptre     (b) -1 dioptre        (c) 1 dioptre (d)  -100 dioptre

41.   All distances of the lens measure from the _____

(a) Pole                (b) Centre    (c) None of these   (d) optical Centre

42.   Which of the following is not behaves like a lens.

(a) water     (b) mirror    (c) glass          (d)  both (a) & (b)

43.   Which mirror is used in solar furnaces?

(a) Convex   (b) Concave (c) Plane      (d)  bifocal

44.   Which lens is used in magnifying glass?

(a) Concave (b) Plane      (c) Convex   (d) bifocal

45.   The refractive index of water is:

(a) 1.3                 (b) $3 \times 10^8$ m/s      (c) 1.00033   (d) 1.52

46.     Which of the following is the lens formula?

       (a) $1/v - 1/u = f$            (b) $1/v - 1/u = 2/R$

       (c) $1/R - 1/u = 2v$           (d) $1/v - 1/u = R/f$

47. Refraction can take place in?

       (a) convex mirror   (b) glass slab        (c) concave mirror (d) None of these

48.     In a glass slab. The incident ray is _____ to the emergent ray:

       (a) Same      (b) equal     (c) parallel   (d) collinear

49.     A Girl saw her in a magic mirror she saw the upper part in same middle is enlarged and the lower is diminished in the mirror. Then what types of mirrors are:

       (a) Convex, Concave and Plane          (b) Plane, Convex and Concave

       (c) Plane, Concave and Convex         (d) Concave, Plane and Convex

**ANSWERS:**

| QUE. | ANS. | QUE. | ANS. | QUE. | ANS. | QUE. | ANS. | QUE. | ANS. |
|------|------|------|------|------|------|------|------|------|------|
| 1 | D | 11 | C | 21 | B | 31 | A | 41 | D |
| 2 | D | 12 | B | 22 | B | 32 | B | 42 | B |
| 3 | A | 13 | C | 23 | B | 33 | A | 43 | B |
| 4 | A | 14 | B | 24 | B | 34 | A | 44 | C |
| 5 | B | 15 | B | 25 | B | 35 | B | 45 | C |
| 6 | C | 16 | B | 26 | B | 36 | B | 46 | B |
| 7 | D | 17 | B | 27 | B | 37 | C | 47 | B |
| 8 | B | 18 | B | 28 | B | 38 | B | 48 | C |
| 9 | B | 19 | C | 29 | A | 39 | D | 49 | C |
| 10 | B | 20 | D | 30 | C | 40 | B | | |

# 7. HUMAN EYE

## SOME IMPORTANT POINTS

- Light enters in the human eye through cornea.
- The eye ball is 2.3 in diameter.
- Iris regulates the size of pupil.
- Pupil regulates the amount of light entering in the eye.
- The crystalline lens focuses the image on the retina.
- The lens is biconvex lens.
- Retina is light sensitive screen.
- Retina generates electric signal and these signals sent via optic nerve.
- The adjustment of focal length of crystalline lens with the help of cillary muscle called power of accommodation.
- The near point of normal vision is about 25cm.
- The far point of normal vision is infinity.
- Myopia: In this defect eye can see near by object clearly but cannot see distant object distinctly.
- The image formed in this defect front of retina.
- Causes of myopia elongation of eye ball :Decrease in focal lens
- This defect is corrected by using a concave lens.
- Hypermeteropia: A person can see distant object distinctly but cannot see near by object clearly.
- The image formed behind the retina.
- Causes of hypermeter 1. Eye ball become too smalt 2. Increase in focal length.
- This defect is corrected by using a convex lens.
- Presbyopia: As we become older we cannot see near and distant object clearly.
- Causes:
  1. Gradually weakening  of a cilliary muscle
  2. Flexibility of eye lens decreases.

- It is correct by bifocal lens upper past consist concave lens and lower past convex lens.
- Prism splits white light into a band of colours.
- These components of white light are called its spectrum.
- The splitting of white light in to its colour components is called dispersion.
- The red light band least violet band the most.
- In rainbow formation first dispersion and refraction of light takes place and internal after this again dispersion and reflection takes place.
- Twinkling of star apparent position of star advance sunrise and delayed sunset all phenomena are happening due to atmospheric refraction.
- The scattering of light by colloidal particle is called tyndall of effect
- The colour of sun at sunrise and sunset all happens due to scattering of light.

# 7. THE HUMAN EYE

1.  We can identity objects by a sense organ called?

    (a) nose   (b)eye   (c)tongue   (d)none of these

2.  The image formed on a light sensitive screen is called?

    (a)retina   (b)cornea   (c)iris   (d)none of these

3.  Light enters the eye through a membrane called?

    (a)iris   (b)cornea   (c)retina   (d)none of these

4.  The diameter of eyeball is approx?

    (a)2.3cm   (b)2.5cm   (c)3cm   (d)none of these

5.  A dark muscular diaphragm which controls the size of pupil is?

    (a)iris   (b)pupil (c)cornea   (d)none of these

6.  The amount of light entering the eye is regulated by?

    (a) iris   (b)pupil   (c)cornea   (d)none of these

7.  The adjustment of focal length required for objects at different distances is provided by?

    (a)iris   (b)crystalline lens   (c)eye ball   (d)none of these

8.  The image formed on retina is?

    a) real and inverted   (b)virtual   (c)erect   (d)none of these

9.  The electrical signals generated by light sensitive calls are sent to brain via?

    (a) Cornea   (b) retina   (c) optic nerve   (d) none of these

10. In the dim light iris expands the?

    (a) Retina   (b) pupil   (c) cornea   (d)none of these

11. The visual impairment is due to damage of?

(a) Retina or optic nerve  (b)iris  or pupil

(c) Cornea or eye lens     (d) all of these

12. We can see distant objects clearly by increasing in?

(a) Curvature  (b) focal length  (c) pole  (d)none of these

13. We can see nearby objects by increasing in?

(a) Curvature    (b) focal length  (c)pole  (d)none of these

14. The ability of eye lens to adjust its focal length is called?

(a)refraction  (b)magnification  (c)accommodation (d)none of these

15. The near point of view of a normal eye is?

(a)20 cm  (b)25cm   (c)30  cm   (d)none of these

16. The far point of view of a normal eye is?

(a) Infinity   (b) 40 cm  (c) 25cm  (d)none of these

17. The crystalline lens of people milky and cloudy this condition is?

(a)myopia  (b)cataract  (c)magnification   (d)none of these

18. A person can see nearby object clearly but cannot see distant objects distinct by due to?

(a)myopia  (b)presbyopia  (c)hypermetropia  (d)none of these

19. The defect which arise due to excessive curvature of the eye lens is

(a)myopia (b)presbyopia  (c)hypermetropia  (d)none of these

20. Myopia can be corrected by using?

(a) Concave lens  (b) convex lens (c)bifocal lens (d)none of these

21. A person can see distant objects clearly but cannot see near objects distinctly due to?

    (a)myopia (b)hypermetropia (c)presbyopia (d)none of these

22. The defect which arises due to increasing in focal length is?

    (a)myopia (b)presbyopia (c)hypermetropia (d)none of these

23. In a myopic eye the image of object is formed on?

    (a) retina (b)front of retina (c)behind of retina (d)none of these

24. In a hypermetropic eye the image of object is formed on?

    (a)retina ( b)front of retina (c)behind of retina (d)none of these

25. Hypermetropia can be corrected by image of object is formed on?

    (a) retina (b)front of retina (c)behind of retina (d)none of these

26. A bi focal lens consists of?

    (a) convex lens (b)concave lens (c)both (a)and (b) (d)none of these

27. The defect which arises due to weakening of the cilliary muscles is?
    (a)myopia                      (b)hypermetropia (c)presbyopia
    (d)none of these

28. The upper portion of a bi –focal lens consist?

    (a) concave lens (b)convex lens

    (c) both (a)and (b) (d)none of these

29. The lower portion of bi –focal lens is consisting of?

    (a)concave lens (b)convex lens (c)both (a)and (b) (c)none of these

30. It is possible to correct refractive defects by using?

(a)contact lens          (b)surgical intervention

(c)both (a)and (b)      (c)none of these

31. in a glass which ray is parallel to the incident ray during refraction?

(a) normal ray   (b)refracted ray  (c)emergent ray  (d)none of these

32. The number of rectangular lateral surfaces in a glass prism is?

(a)four  (b)three  (c)two  (d)none of these

33. The number of triangular bases in a glass prism?

(a)four  (b)three  (c)two  (d)none of these

34. The angle between two lateral surfaces of a prism ?

(a)angle of emergences  (b)angle of prism

(c)angle of incidence    (d)none of these

35. When a light ray entering to the denser medium form rarer medium it bends towards?

(a)normal  (b)incident ray (c)emergent ray  (d)none of these

36. The spectrum of light means?

(a)band of coloured components  (b)group of colures
(c)both  (a)and (b)                  (d)none of these

37. The various colures present in sunlight are?

(a)violet, indigo, blue, green, yellow, orange, red

(b) violet , indigo , brown , green, yellow ,orange, red

(c) violet , indigo , brown ,pink , yellow , orange

(D) none of these

38. The splitting of light into its colored components is called?

    (a) scattering  (b)dispersion  (c)spectrum  (d)none of these

39. The colour which bends the least during dispersion of light is?

    (a)violet (b)green  (c)red  (d)none of these

40. Who was the first to use a glass prism to obtain the spectrum of sunlight?

    (a)gallileo  (b)Issac Newton  (c)Archimedes   (d)none of these

41. Rainbow of light formed due to?

    (a)scattering of light          (b)dispersion of light

    (c)refraction of light          (d)none of these

42. A rainbow is always formed in the dire of?

    (a) infront of the sun   (b)opposite the sun  (c)both (a)and (b) (d)none of these

43. The different colours of rainbow reach to observer's eye due to?

    (a) dispersion of light     (b total internal reflection

     (c) both (a)and (b)         (d)none of these

44. The hotter air present in atmosphere is?

    (a) lighter that cooler air   (b)heavier than cooler air

     (c)both (a) and (b          (d)none of these

45. The twinkling of stars is due to?

    (a) scattering              (b)atmospheric refraction

    (c) dispersion of light    (d)none of these

46. The time difference between sunset and the apparent sunset is about ?

    (a)5 minutes  (b)2 minutes  (c)10 minutes  (d)none of these

47. The path of a beam of light become visible when it through?

    (a) A true solution        (b) suspension solution

    (c) Colloidal solution      (d)none of these

48. If the size of scattering particles is large enough the scattered light

    may appear?

    (a)red  (b)white  (c)violet  (d)none of these

49. The colour of sky appear blue due to?

    (a)scattering of light              (b)dispersion of light

    (c)atmospheric refraction           (d)none of these

50. The colour which has the longest wavelength is?
    (a) red            (b)violet            (c)green            (d)blue

**ANSWERS:**

| QUES. | ANS. | QUES. | ANS. | QUES. | ANS. | QUES. | ANS. | QUES. | ANS. |
|-------|------|-------|------|-------|------|-------|------|-------|------|
| 1     | B    | 11    | D    | 21    | B    | 31    | C    | 41    | B    |
| 2     | A    | 12    | B    | 22    | C    | 32    | B    | 42    | B    |
| 3     | B    | 13    | A    | 23    | B    | 33    | C    | 43    | C    |
| 4     | A    | 14    | C    | 24    | C    | 34    | B    | 44    | A    |
| 5     | C    | 15    | B    | 25    | B    | 35    | A    | 45    | B    |
| 6     | B    | 16    | A    | 26    | C    | 36    | C    | 46    | B    |
| 7     | B    | 17    | B    | 27    | C    | 37    | A    | 47    | C    |
| 8     | A    | 18    | A    | 28    | A    | 38    | B    | 48    | B    |
| 9     | C    | 19    | A    | 29    | B    | 39    | C    | 49    | A    |
| 10    | B    | 20    | A    | 30    | C    | 40    | B    | 50    | A    |

# ELECTRICITY

## SOME IMPORTANT POINTS

- The rate of flow of electric charge is known as electric current.
- Electric charge is denoted by "Q" and its SI unit is coulomb.
- When a net charge Q flows in any cross section conductor in time T then current I is I = Q/T (I = current, Q = charge ).
- The SI unit of current in Ampere.
- 1Coulomb in equivalent to charge contained $6.25*10^{18}$ electrons.
- 1Electron passes negative charge is usually is nearly $1.6*10^{-19}$ coulomb.
- A continuous and closed path of an electric current is called an electric circuit.
- Electric current is measured by a device called Ammeter.
- It is always connected in series in a circuit.
- An ideal Ammeter has low resistance.
- The electron moves only if there is a difference of electric pressure called the potential difference along the conductor.
- The electric potential difference is defined as difference between two points in an electric circuit carrying some current as the work done to move a unit charge form one point to the other.
- Potential difference is denoted by "V" and its SI unit is volt. V= W/Q, 1V = J/C.
- Potential difference is measured by an instrument called voltmeter and it is always connected in parallel in a circuit.
- It is high resistance.
- Ohm`s law stated as electric current is flowing through a circuit is directly proportional to potential difference across it ends provided its temperature remains the same. V α I, V= IR.
- Where R is constant and it is called resistance. Resistance is a property that resist the flow of electrons in a conductor and its SI unit is ohm 'Ω. R= V/I.
- A component used in a circuit to regulate the electric current with charging voltage source called rheostat.
- Resistance of conductor depends upon :
  1. Length of wire.

2. Area of cross section.
3. The nature of its material.
4. Temperature.

➤ R $\alpha$ l,        R $\alpha$ l/a,      R $\alpha$ l/a,      R = $\rho$ l/a
➤ Where $\rho$ is proportionality constant called 'rho' and it is called resistivity of material.

   The SI unit of resistivity is ohmmeter ($\Omega$m).

➤ Tungsten is used in filament of electric bulbs.
➤ Resistors connected in two ways series and parallel.
➤ In series the equivalent resistance is given by Req = R1+R2+R3.....
➤ In parallel the equivalent resistance is given by 1/Req = 1/R1 + 1/R2 + 1/R3......
➤ Joule law of heating H = $I^2$ RT, H = VIT where H is electric energy electric power is H/T , P = $I^2$R and P = $V^2$/R, P = VI.
➤ The rate at which electric energy is dissipated or consumed in a electric circuit is called power.
➤ The SI unit of power is watt.
➤ The commercial unit of electric energy is kWh commonly known as unit.
➤ 1kwh = $3.6 \times 10^6$ Joule.

# 8. ELECTRICITY

1.  Which quantity is either positive or negative?

    a. Time     b. Charge     c. Volt     d. Both a and b

2.  1C net charge is equivalent to the charged contained in nearly ...... electrons.

    a. $6.25*10^{18}$     b. $6.25*10^{17}$     c. $6.25*10^{-18}$     d. None of these

3.  Which quantity has SI unit coulomb?

    a. Resistance     b. Current     c. Voltage     d. Charge

4.  Rate of flow of net charge is known as?

    a. Resistance     b. Charge     c. Current     d. Power

5.  The SI unit of current is?

    a. Ampere     b. Second     c. Volt     d. Resistance

6.  Current * time =?

    a. Resistance     b. Charge     c. Volt     d. Resistivity

7.  1 micro Ampere =?

    a. Time     b. Charge     c. Volt     d. Both a and b

8.  In an electric circuit the electric current flow in an ........... direction to the flow of electron conventionally.

    a. same     b. opposite     c. both a and b     d. None of these

9.  Which quantity is measured by ammeter?

    a. Resistance     b. Power     c. Current     d. None of these

10. Ammeter has ............. resistance.

    a. High    b. normal    c. current    d. low

11. In the electric circuit how many cells are there?

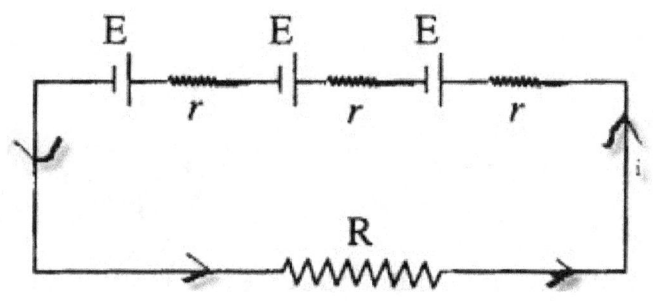
Diagram

    a. 2    b. 3    c. 4    d. 5

12. Voltage *Charge =?

    a. Current    b. Resistance    c. Work    d. None of these

13. Potential difference is denoted by?

    a. W    b. Q    c. V    d. P

14. Which quantity is measured by voltmeter?

    a. work    b. Charge    c. Potential difference    d. Power

15. Which of the following is scalar quantity?

    a. Current    b. Potential Difference    c. Charge    d. All of these

16. Voltmeter has ........... resistance.

    a. High    b. Normal    c. Both a and B    d. low

17. Which is always connected in parallel across the circuit?

    a. Ohmmeter    b. Ammeter    c. Speedometer    d. voltmeter

18. What is the symbol of joint wire?

a.

b.

c.

d.  none of these

19. Who found the relationship between the current (I) flowing through a conductor and potential difference (V) across the terminals of a conductor using circuit diagram.

a. George Simon Ohm     b. Andrew Marie Ampere

c. Glenn Maxwell     d. None of these

20. Resistance =?

a. Voltage *Current     b. Voltage/Current

c. Current/Voltage     d. All of these

21. The property of a conductor that opposes the flow of current is known as?

a. Resistance     b. Rheostat     c. Voltage     d. None of these

22. What is the SI unit of resistance?

a. Volt     b. Ohm     c. Watt     d. None of these

23. Which of the following is a resistor?

a. Human     b. Wire     c. Wood     d. All of these

24. The resistance of a conductor depends on?

a. temperature   b. Volume of material  c. both a and b    d. None of these

25.    What is the SI unit of resistivity?

    a. Kilowatt-hour     b. ohm-meter       c. ohm    d. volt

26.    Which element is used almost exclusively for filament of electric bulb?

    a. Titanium     b. Copper     c. Tungsten   d. alluminium

27.    Which elements are generally used for electrical transmission lines?

    a. Alluminium     b. Copper     c. Both a and b   d. None of these

28.    Which of the following is a series combination?

    a.

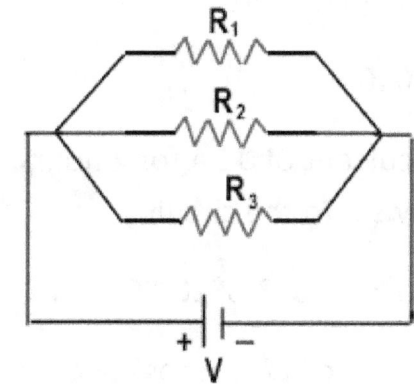

    b.

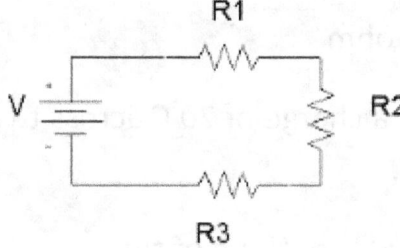

    c. Both a and b

    d. None of these

29.    Which quantity remains same in combination of resistors?

a. Potential difference     b. Current     c. Charge     d. Time

30.    Which gases are filled in bulb to prolong the life of filament?

a. Nitrogen and Helium     b. Helium and Argon

c. Argon and Oxygen     d. Nitrogen and Argon

31.    1 Volt * 1 Ampere =?

a. 1C     b. 1J     c. 1W     d. 1A

32.    A continuous and closed path of an electric current is known as?

a. Electric circuit     b. Electric current     c. Electric rode     d. None of these

33.    A current of 0.2 A is drawn by an electric bulb for 22 minutes. Find the amount of electric charge that flow through the circuit.

a. 2640 A     b. 264 C     c. 264 A     d. 2640 C

34.    The filament of an electric CFL draws a current of 0.5A for 2 hours. Calculate the amount of charge that flows into the circuit.

a. $3.6*10^2$ C     b. $3.6*10^{-2}$ C     c. $3.6*10^3$ C     d. $3.6*10^{-3}$ C

35.    Calculate the work done in moving a charge of 12 C across two points having a potential difference of 7 V.

a. 84 J     b. 84 W     c. 84 C     d. 84 ohm

36.    Calculate the work done in moving a charge of 20 C across two points having a potential difference of 5V.

a. $8*100$ J     b. $8^0*100$ J     c. $8*1000$     d. None of these

37.    Given R = 440 ohm, V = 220 V, then I=?

a. 0.5 A     b. 0.05 A     c. 5 A     d.  All of these

38.    Given, V = 220 V, R = 600 ohm, I =?

a. 0.003     b. 0.3     c. 0.0003     d.  0.03

39.    Given V = 100 V, I = 5 A, R =?

a.  20 ohm-m     b. 20 ohm     c. 200 ohm-m     d.  200ohm

40.    3 resistor connected in parallel combination of resistance 10ohm, 20 ohm, 25 ohm. What is its equivalent resistance?

a. 0.019 ohm     b. 0.19 ohm     c. 100/9 ohm     d.  19/100 ohm

41.    Given P = 840 W, t = 10s, W = ?

a. 8400 J     b. 8400 C     c. 8.4 J     d.  None of these

42.    Given I = 10 A, R = 10 ohm, t = 0.5 s, H = ?

a. 8400 J     b. 8400 C     c. 8.4 J     d.  None of these

43.     An electric bulb is connected to a 220 v battery. The current is 3 A. what is the power of bulb?

a. 6.6 W    b. 66 W     c. 660 W     d.  All of these

44.    A CFL is rated at 5V, 100A. What is its power?

a. 500 W     b. $5^0$*100 W     c. 50*10 W     d.  Both a and c

45.    An electrical fan rated 200W operates 4 hours per day. What is the cost of the energy to operate it for 30 days at the rate of Rs.2 per KWh?

a. 28     b. 40     c. 288    d.  None of these

46.    A charge of 700 C flowing in 10 s.  How much current is flowing through the circuit?

a. 70 A     b. 70 J     c. 70 W     d.  None of these

47.    Calculate the work done in moving a charge of 120 C across two points having a potential difference of 5 V?

a. 500 J    b. 600 J    c. 525 J    d. None of these

48.    H = 25 J, I = 5A, t = 1s , R = ?

a. 1 ohm    b. 0.1 ohm    c. 0.01 ohm    d. 0.001 ohm

49.    An electrical heater takes 10 A current from a 220 V line. Determine power of heater and energy consumed in 5 h.

a. 2200, $3.96*10^8$ J    b. 2200, $3.96*10^7$ J

c. 2020, $3.96*10^7$ J    d. 2000, $3.96*10^7$J

ANSWERS:

| QUE. | ANS. | QUE. | ANS. | QUE. | ANS. | QUE. | ANS. | QUE. | ANS. |
|------|------|------|------|------|------|------|------|------|------|
| 1 | D | 11 | B | 21 | A | 31 | C | 41 | A |
| 2 | A | 12 | C | 22 | B | 32 | A | 42 | B |
| 3 | D | 13 | C | 23 | D | 33 | B | 43 | C |
| 4 | C | 14 | C | 24 | A | 34 | C | 44 | D |
| 5 | A | 15 | D | 25 | B | 35 | A | 45 | D |
| 6 | B | 16 | A | 26 | C | 36 | B | 46 | A |
| 7 | C | 17 | D | 27 | C | 37 | A | 47 | B |
| 8 | D | 18 | A | 28 | B | 38 | D | 48 | A |
| 9 | C | 19 | A | 29 | B | 39 | B | 49 | B |
| 10 | D | 20 | B | 30 | D | 40 | C | | |

# 9. MAGNETIC EFFECT OF ELECTRIC CURRENT

## SOME IMPORTANT POINTS

- The wire carrying electric current behaves like a magnetic.
- The area in which force of attraction & repulsion is felled called magnetic field
- Magnetic field cannot intersect each other
- Magnetic field denser at poles.
- Magnetic field originates from North Pole to South pole in bas magnet.
- Load stone is natural magnet called Hematite.
- Magnetic field is rector quantity.
- Magnitude of magnetic field is directly proportional to current.
- Right hand thumb rule stated as if your thumb to be the direction of current then your wrapped fingers show the direction of magnetic field lines.
- In a circular coil having turns the field produced times.
- Field lines inside the solenoid is parallel and it behaves like a bar magnet.
- Fleming's left hand rule: If the fore finger shows the direction of field lines and the middle finger show direction of current then thumb shows the direction of motion
- The principle of electric motor is Fleming's left hand rule and electric motor convert electrical energy in to mechanical energy.
- The principle of generator based upon electromagnetic Induction.
- The process by which a changing of magnetic field in a conductor induced current is called electromagnetic induction.
- In generator mechanical energy converted into electrical energy.
- The time varying current is known as alternative current.
- The current does not its direction with time called direct current
- Ac changes direction after every 1/100 second in India and frequency of AC is 50 $H_z$
- In home all appliances connected in parallel.
- Fuse is a safety device of all domestic circuits.

- Fuse is made up of alloy of copper and tin.
- In our houses we get AC of 220 volt.

# 9. MAGNETIC EFFECT OF ELECTRIC CURRENT

1.  An electric current – carrying conductor wire behaves like a?

    a. magnet      b. conductor      c. galvanometer d. motor

2.  The electric current through the copper wire has produced a:

    a. magnetic effect     b. condutivity   c. tindal effect   d. none of these

3.  Magnet has properties of........and.........

    a. attraction b. both (a) and (c)     c. repulsion     d.none of these

4.  Magnets consist of a number of oxides of iron with......... its structure

    formula:

    a.  $Fe_3O_4$     b. $co_2$     c.$Fe_4O_3$          d.$FE_4O_3$

5.  Natural magnets are:

    a.Strong     b. weak     c.     irregular In shape          d.both (a)and (b)

6.  The chemical properties of magnet and iron is:

    a.opposite     b.same     c.same but different     d. none of these

7.  A compass needle gets deflects when we brought it........

    a.near a bar megnet               b.far from a bar megnet

    c. carry a bar megnet             d. none of these

8.  A bar magnet is a:

    a. natural magnet     b.electromagnet

    c. artificial magnet       d. none of these

9.  There are two ends of magnets:

    a. south- north                b. southern east- north

c. northern east- west          c. east-west

10.   When an iron and magnet comes into contact:

a. attract each other          b. repel each other

d. crunch each other          d.all of above

11.   .........poles of magnet attract each other.

a. unlike     b. both (a)and (c)   c. like        d. none of these

12.   Choose the incorrect statement:

a. magnet field lines are parallel and equidistant.

b. $Fe_3O_4$ is the structural formula of magnet.

c. magnetic field lines are closed and curve.

d. both a and c.

13.   What is the SI unit of magnetic field line?

a. ampere     b. torque     c. Newton       d. tesla

14.   The Fleming's left hand rule applies on?

a. motor     b. generator   c. T.V       d. Bike

15.   choose the correct option:

a. the magnet exerts its influence in its surroundings

b. magnetic field lines form closed loop.

c. in presence of magnetic field the iron fillings are arranged in a pattern.

d. all of these

16.   When current flows magnetic field produced is ............... to the direction of the current flow.

a. parallel    b. perpendicular    c. straight    d. none of these

17.    When the magnitude of the current increases the magnitude of the magnetic field ..........?

a. increases    b. decreases    c. first increases then decreases    d. None

18. If magnitude of magnetic field increases then deflection in compass.....?

a. increases    b. decreases    c. remains constant    d. both a and c

19.    The magnetic field lines emerges from?

a. south pole    b. north pole    c. either south or north    d. none

20.    The magnetic field lines merge at?

a. north pole    b. south pole    c. either south or north    d. none

21.    Inside the magnet the magnetic field direction is from?

a. north to south    b. north to east    c. south to east    d. south to north

22.    Magnetic field is?

a. vector    b. scalar    c. Constant    d. none

23.    The field lines are stronger when we keep the magnets.

a. closer    b. far    c. field lines are same everywhere    d. none

24.    Choose the correct option.

a. magnetic field lines merges at south  pole

b. magnetic field lines are invisible

c. like pole repel each other

d. all of these

25.    Choose the correct option.

a. no two field lines cross each other

b. Two field lines crosses each other

c. magnetic field lines are parallel to each other

d. none of these

26. An electric current through a metallic conductor produces:

a. electric field     b. cooling     c. energy     d. magnetic field

27. If the current flows from north to south the north pole of the compass needle would move towards the?

a. west     b. east     c. north     d. south

28. The compass deflects due to:

a. magnetic field     b. electric field     c. field lines     d. none

29. In which element electric field lines surround it?

a. conductor     b. insulator     c. semiconductor     d. none

30. Magnetic field lines are ……………to the electric field.

a. parallel     b. perpendicular     c. intersecting     d. none

31. Magnetic field lines are generated by current through a:

a. conductor     b. insulator     c. semiconductor     d. none

32. If we reverse the direction of electric current, the direction of magnetic field will be?

a. reverse     b. forward     c. no change     d. none

33. If current increases:

a. deflection in compass increases

b. deflection in compass decreases

c. remains constant

d. none

34. Right hand thumb rule indicates:

a. magnetic field lines are parallel to electric current

b. magnetic field lines are perpendicular to electric current

c. magnetic field lines are bisectors of electric field lines

d. none

35. The magnetic field produced by current carrying straight wire depends on the distance from it, as

a. inversely    b. directly    c. 4 times    d. none

36. By time we reach the centre of the circular loop, the arc of the big circle would appear as:

a. straight line    b. curved line    c. intersecting line    d. none

37. To check whether the every section of the wire contribute to the magnetic field lines in the same direction within the loop, we apply:

a. right hand rule    b. left hand rule    c. both a and b    d. none

38. Every point on the wire carrying current would give rise to the magnetic field appearing as straight line at the centre of the:

a. loop    b. motor    c. tube light    d. generator

39. Loop is a:

a. conductor    b. semiconductor    c. insulator    d. None

40. Magnetic field is the:

a. force    b. energy    c. both a and b    d. none

41. A coil produces magnetic field:

    a. when electrons flow                  b. when electrons becomes static

    c. due to magnetic force of coil        d. none

42. The current in each circular turn has the;

    a. same direction    b. opposite direction

    c. parallel direction of coil    d. none

43. A coil of many circular turns of insulated copper wire wrapped closely in the shape of cylinder is called a:

    a. loop    b. coil    c. none of these    d. solenoid

44. The pattern of magnetic field lines of the solenoid is same as pattern of:

    a. bar magnet    b. conductor    c. loop    d. coil

45. One end of solenoid behaves as North Pole; second end of the solenoid behaves as:

    a. south pole    b. north pole    c. east pole    d. none

46. Soft iron when placed inside the coil then it changed to:

    a. electromagnet    b. conductor    c. plastics    d. None of these

47. The magnetic field inside a long solenoid carrying current:

    a. is zero    b. decreases as we move towards its ends

    c. increases as we move towards its ends    d. is same at all points

48. The magnet must exert an equal and opposite force on current carrying:

    a. conductor    b. insulator    c. semiconductor    d. None of these

49. The device used for producing electric current is:

a. Galvanometer    b. Remote    c. Generator    d. Motor

50. The direction of the force on the conductor depends upon the direction of:

a. current and magnetic field    b. magnetic and electric field

c. current and force    d. None of these

51. An electric motor converts electrical energy into:

a. Mechanical energy    b. potential energy    c. kinetic energy    d. none

52. The energy stored in cell which produces electrical energy is:

a. south pole    b. north pole    c. east pole    d. none

53. Flow of charge due to varying magnetic field with respect to the conductor is called:

a. magnetic flux    b. electromagnetic induction

c. magnetic field    d. None of these

54. The direction of the magnetic field is perpendicular to the current motion. It experience ............... for moving a coil.

a. pressure    b. thrust    c. force    d. none

55. In an electric generator the ............ is used to rotate a conductor in a magnetic field to produce electricity.

a. mechanical energy    b. potential energy    c. kinetic energy    d. none

ANSWERS:

| QUE. | ANS. | QUE. | ANS. | QUE. | ANS. | QUE. | ANS. | QUE. | ANS. | QUE. | ANS. |
|------|------|------|------|------|------|------|------|------|------|------|------|
| 1 | A | 11 | A | 21 | D | 31 | A | 41 | A | 51 | A |
| 2 | A | 12 | A | 22 | A | 32 | A | 42 | A | 52 | D |
| 3 | B | 13 | D | 23 | A | 33 | A | 43 | D | 53 | B |
| 4 | A | 14 | A | 24 | D | 34 | B | 44 | A | 54 | C |
| 5 | D | 15 | D | 25 | A | 35 | A | 45 | A | 55 | A |
| 6 | A | 16 | B | 26 | D | 36 | A | 46 | A | | |
| 7 | A | 17 | A | 27 | B | 37 | A | 47 | D | | |
| 8 | A | 18 | A | 28 | A | 38 | A | 48 | A | | |
| 9 | A | 19 | B | 29 | A | 39 | A | 49 | C | | |
| 10 | A | 20 | A | 30 | B | 40 | A | 50 | A | | |

# NOTES

www.ingramcontent.com/pod-product-compliance
Lightning Source LLC
Chambersburg PA
CBHW080829180526
45168CB00006B/2624